Now I Know

What's Under the Ocean

Written by Janet Craig
Illustrated by Paul Harvey

Troll Associates

Library of Congress Cataloging in Publication Data

Craig, Janet.
    What's under the ocean.

    (Now I know)
    Summary: Brief text and pictures introduce some
animals and plants that live in the ocean.
    1. Marine biology—Juvenile literature.
[1. Marine biology.   2. Ocean]   I. Harvey, Paul,
1926-   , ill.   II. Title.
QH91.16.C7   574.92   81-11425
ISBN 0-89375-652-0   AACR2
ISBN 0-89375-653-9 (pbk.)

10   9   8   7   6   5   4   3   2   1

Look at the ocean!

It is very big. It is very deep.

Who lives here?

Who lives beneath the waves?

Many things live in the ocean.

Some are large and some are small.

Some come in beautiful, bright colors.

Look who's here—a shark.

Here is a giant sea turtle.

Here comes the biggest animal in the ocean—

the whale.

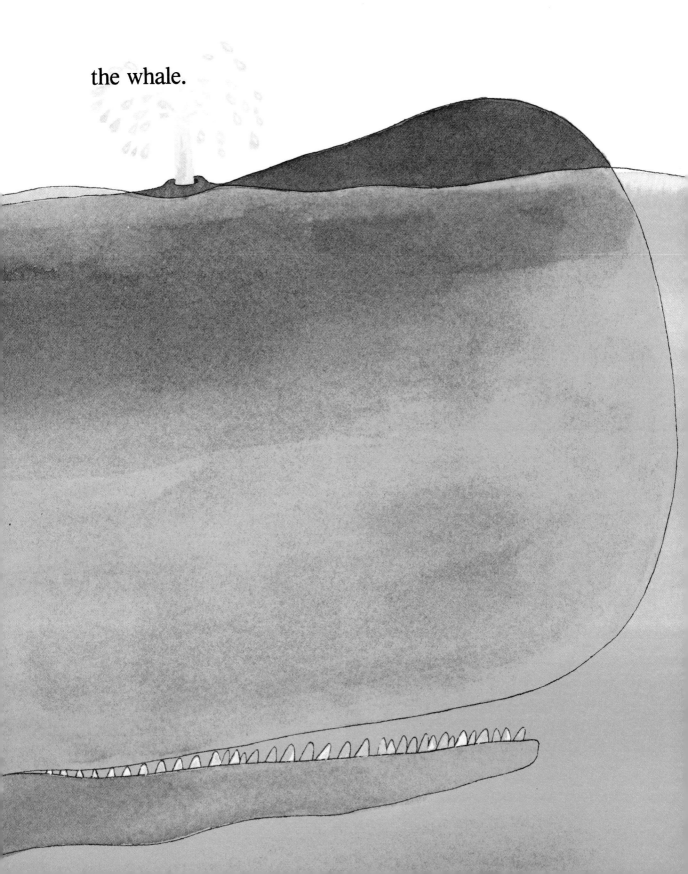

Here comes one of the smallest animals in the ocean—

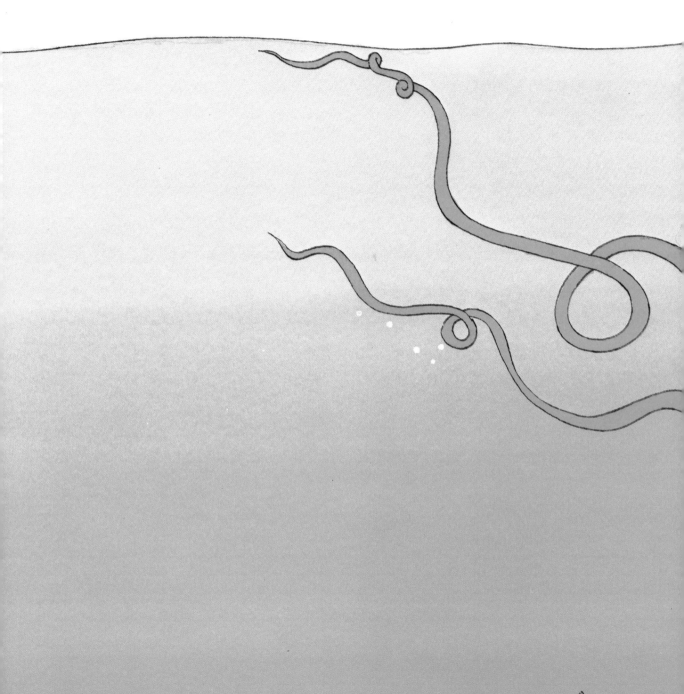

a tiny sea horse.

A baby dolphin plays with her mother.

They jump up. They dive down.

They call to one another.

"Click, clickety, click," says the mother to her baby.

"Let's go deeper."

Does anyone live way down here?

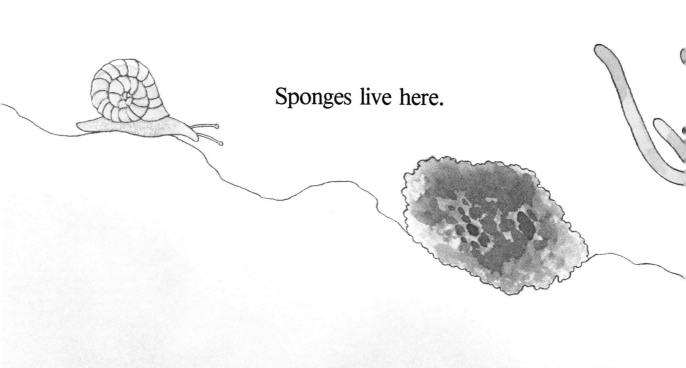

Sponges live here.

Corals live here, too. What else?

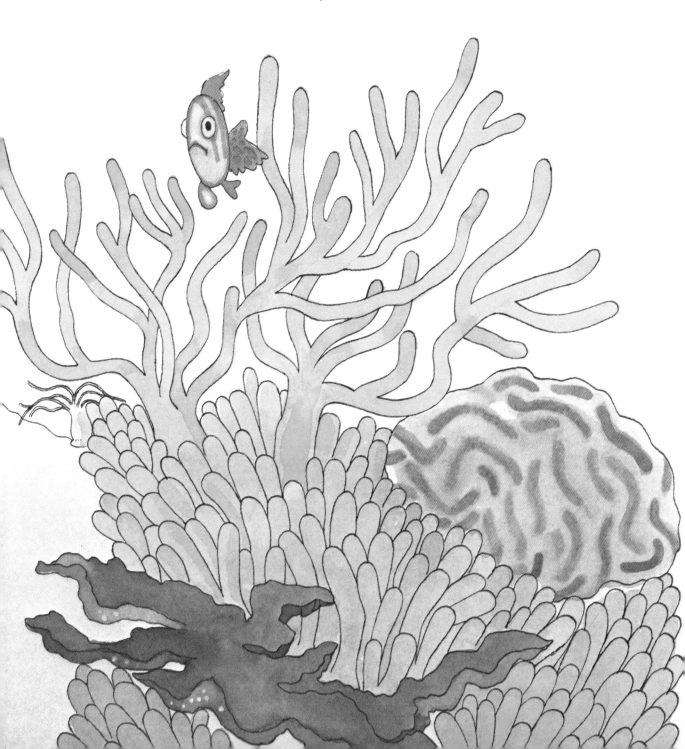

The crab lives on the sandy bottom.

Don't get in his way! He'll pinch you if you do.

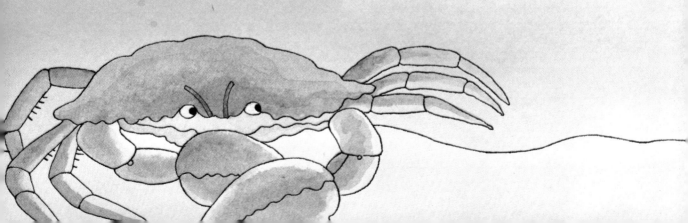

Seaweed grows in the ocean.
Some seaweeds are so small, you
cannot see them. Others are very,
very long.

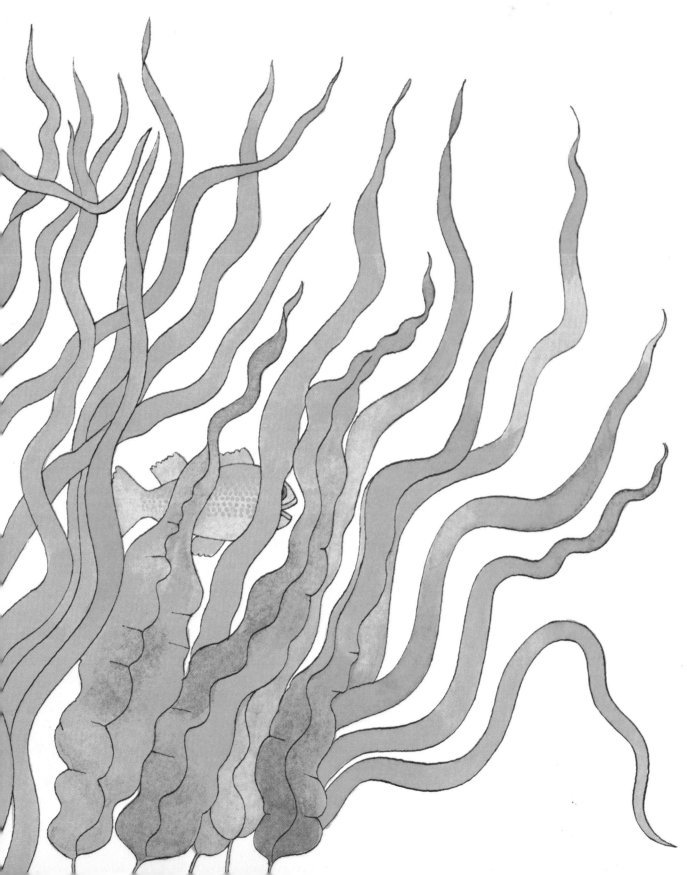

Many, many animals and plants live under the ocean.

The ocean is their home.